READING POWER

Man-Made Disasters

Burning Up

Losing Our Ozone Layer

August Greeley

The Rosen Publishing Group's
PowerKids Press™
New York

Published in 2003 by The Rosen Publishing Group, Inc.
29 East 21st Street, New York, NY 10010

Copyright © 2003 by The Rosen Publishing Group, Inc.

All rights reserved. No part of this book may be reproduced in any form without permission in writing from the publisher, except by a reviewer.

First Edition

Book Design: Christopher Logan

Photo Credits: Cover © Premium Stock/Corbis; pp. 4–5, globe logo pp. 6, 10, 14, 17 © PhotoDisc; p.6 © Dean Conger/Corbis; p. 7 © Bill Ross/Corbis; pp. 7, 10, 11, 16, 21 (map and illustrations) Chris Logan; pp. 8–9 © Joseph Sohm/ChromoSohm Inc./Corbis; pp. 11, 20 © Ecoscene/Corbis; pp. 12, 15 (inset) © AP/Wide World Photos; p. 13 © Photri Microstock, Inc.; pp. 14–15 © John Noble/Corbis; p. 17 © Reuters NewMedia Inc./Corbis; p. 18 © Lowell Georgia/Corbis; p. 19 © Philip Gould/Corbis; p. 21 © Morton Beebe/Corbis

Library of Congress Cataloging-in-Publication Data

Greeley, August.
  Burning up : losing our ozone layer / August Greeley.
    p. cm. — (Man-made disasters)
  Summary: Introduces what the ozone layer is, how it protects life on Earth, the damage done to it by CFCs, the danger of the increasingly large hole in the ozone layer, and what may be done to heal it.
  Includes bibliographical references and index.
  ISBN 0-8239-6482-5 (lib. bdg.)
  1. Ozone layer depletion—Juvenile literature. [1. Ozone layer. 2. Ozone layer depletion.] I. Title.
  QC879.712 .G47 2003
  363.738'75—dc21
                                                    2002000526

Contents

Keeping Earth Safe	4
Harmful Chemicals	8
The Ozone Hole	12
Saving the Ozone Layer	16
Glossary	22
Resources	23
Index/Word Count	24
Note	24

Keeping Earth Safe

An important layer of gases floats high above Earth, keeping life in our world safe. This layer is known as the ozone layer.

A layer of gases floats above Earth.

However, the ozone layer is being torn apart by man-made chemicals in the air. It is getting thinner each year. If the ozone layer gets too thin, it won't be able to keep us safe.

Ozone is found naturally 10 to 30 miles above Earth in the stratosphere. This ozone protects Earth from harmful beams of sunlight called ultraviolet (UV) rays. UV rays can cause skin cancer and blindness. They can also hurt crops and even kill small sea animals.

Check It Out

Ozone that forms near the ground is toxic. It makes breathing difficult. It also makes a kind of dirty air called smog.

The stratosphere helps to keep the Sun's harmful UV rays from reaching the surface of the earth.

Harmful Chemicals

Chemicals called CFCs destroy ozone. These chemicals were made to be used in things such as spray cans, refrigerators, and air conditioners. When CFCs were made, it was not known that they would cause a problem.

CFCs aren't harmful to humans, but they cause big problems when they reach the ozone layer.

In 1978, all U.S. companies stopped selling spray cans that used CFCs.

When CFCs float to the stratosphere, the Sun's UV rays break them apart. Chlorine atoms in the CFCs are freed. The chlorine atoms attack and destroy ozone molecules.

Check It Out

A single chlorine atom can tear apart 100,000 ozone molecules.

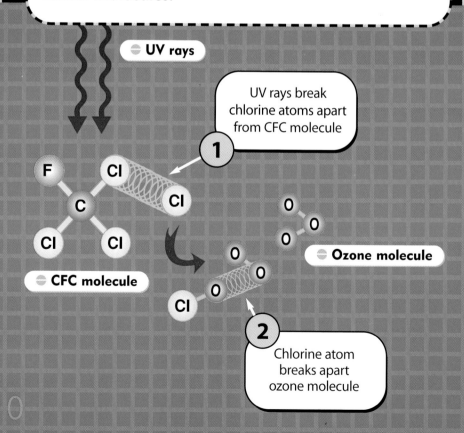

CFCs from broken refrigerators and air conditioners in our homes get into the air and float up to the stratosphere.

The Ozone Hole

In the 1970s, scientists began studying what CFCs do to ozone. Then, they took pictures of the ozone layer from space. These pictures shocked the scientists.

Some scientists work in domed labs to study the ozone layer.

The ozone layer over Antarctica was getting thinner. The ozone layer was so thin over Antarctica that it looked like a hole had formed.

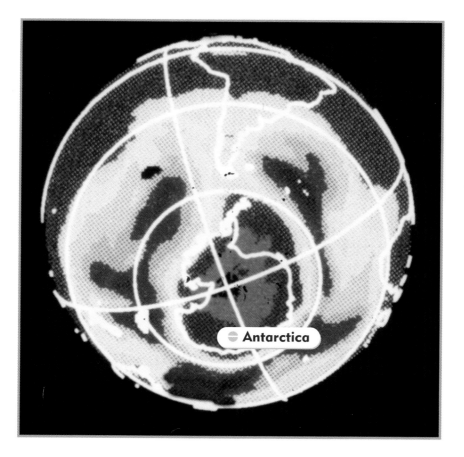

In 1980, this picture of the ozone layer was taken from space. It showed low amounts of ozone over Antarctica. The blue area over Antarctica is the ozone hole.

Scientists learned that the ozone layer over Antarctica is thinnest during September and October. They also discovered that the ozone layer is thinning all around the world.

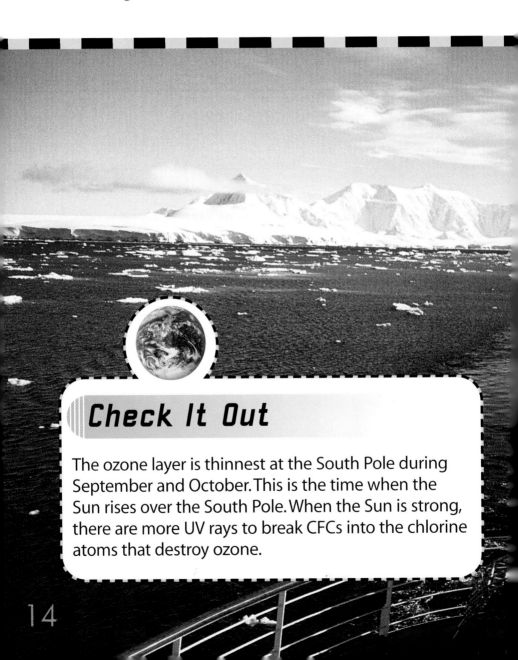

Check It Out

The ozone layer is thinnest at the South Pole during September and October. This is the time when the Sun rises over the South Pole. When the Sun is strong, there are more UV rays to break CFCs into the chlorine atoms that destroy ozone.

This photo was taken above Antarctica in October 1999. It showed that the area with very low amounts of ozone had grown.

During the summer in Antarctica, it is cold enough for large ice caps to stay frozen.

Saving the Ozone Layer

People from all over the world knew they had to find a way to save the ozone layer. In 1987, many nations agreed to follow a plan to stop making and using CFCs and other harmful chemicals that hurt the ozone layer.

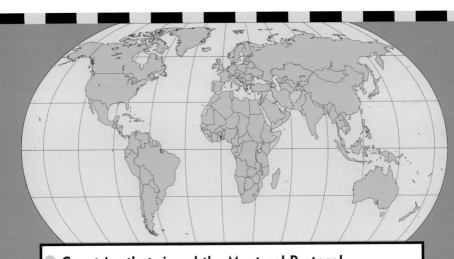

- Countries that signed the Montreal Protocol
- Countries that have not signed the Montreal Protocol

The Montreal Protocol, a plan to save the ozone layer, has been signed by 175 nations.

However, even if every country stopped making CFCs right now, the hole in the ozone layer would remain for many more years.

Check It Out

In 2000, the hole in the ozone layer was larger than ever. It was three times the size of the United States!

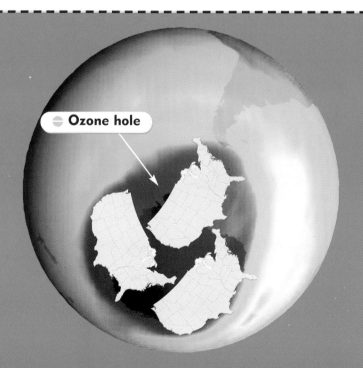

Ozone hole

Governments around the world are making laws to keep the ozone layer safe. CFCs cannot be made or used.

It is important to keep our bodies safe from UV rays. Always wear sunglasses and sunscreen when playing outdoors — even if it is a cloudy day.

Scientists are now making new chemicals that will not destroy the ozone layer.

Scientists have already made new chemicals that can be used in place of CFCs. The new chemicals will not harm the ozone layer.

The hole in the ozone layer is a problem created by humans. Only people can save the ozone layer.

CFCs can be removed from old refrigerators and air conditioners so that they do not get out into the air.

Scientists believe that the ozone hole could be healed by the year 2050 if every nation stops making harmful chemicals like CFCs. That would be good news.

One way to protect the ozone layer is to have your family's car checked to make sure that it is not letting CFCs get into the air.

Glossary

atom (**at**-uhm) one of the tiny pieces that make up all matter

CFCs man-made chemicals that help refrigerators and air conditioners run

chemicals (**kehm**-uh-kuhlz) matter that is made of elements such as carbon and oxygen

chlorine (**klor**-een) a greenish yellow, bad-smelling, poisonous gas

molecule (**mahl**-uh-kyool) a small part of matter that is made of two or more atoms

Montreal Protocol (**mahn**-tree-awl **proh**-tuh-kahl) a plan signed by 175 countries to stop making and using CFCs and other harmful chemicals

ozone (**oh**-zohn) a form of oxygen found naturally in the stratosphere that is a poisonous gas

ozone layer (**oh**-zohn **lay**-uhr) a layer of gases, 10 to 30 miles above Earth, that keeps Earth safe from UV rays

protect (**pruh**-tehkt) to keep safe

skin cancer (**skihn kan**-suhr) a serious illness that hurts the skin and could cause death

smog (**smahg**) a smoky fog caused by pollution

stratosphere (**strat**-uh-sfihr) a layer of gases about 10 to 30 miles above Earth

ultraviolet (UV) rays (**uhl**-truh-**vy**-uh-liht **rayz**) harmful beams of sunlight

Resources

Books

Global Warming: The Threat of Earth's Changing Climate
by Laurence P. Pringle
SeaStar Publishing (2001)

Ozone Hole
by Sally Morgan
Franklin Watts (1999)

Web Sites

Due to the changing nature of Internet links, PowerKids Press has developed an on-line list of Web sites related to the subjects of this book. This site is updated regularly. Please use this link to access the list:

http://www.powerkidslinks.com/mmd/buol/

Index

A
Antarctica, 13–15
atom, 10, 14

C
CFC, 8–12, 14, 16–21
chemicals, 5, 8, 16, 19, 21
chlorine atom, 10, 14

M
molecule, 10

O
ozone, 6, 8, 12–15
ozone layer, 4–5, 8, 12–14, 16–21

S
skin cancer, 6
stratosphere, 6–7, 10–11

U
UV rays, 6–7, 10, 14, 18

Word Count: 485

Note to Librarians, Teachers, and Parents

If reading is a challenge, Reading Power is a solution! Reading Power is perfect for readers who want high-interest subject matter at an accessible reading level. These fact-filled, photo-illustrated books are designed for readers who want straightforward vocabulary, engaging topics, and a manageable reading experience. With clear picture/text correspondence, leveled Reading Power books put the reader in charge. Now readers have the power to get the information they want and the skills they need in a user-friendly format.